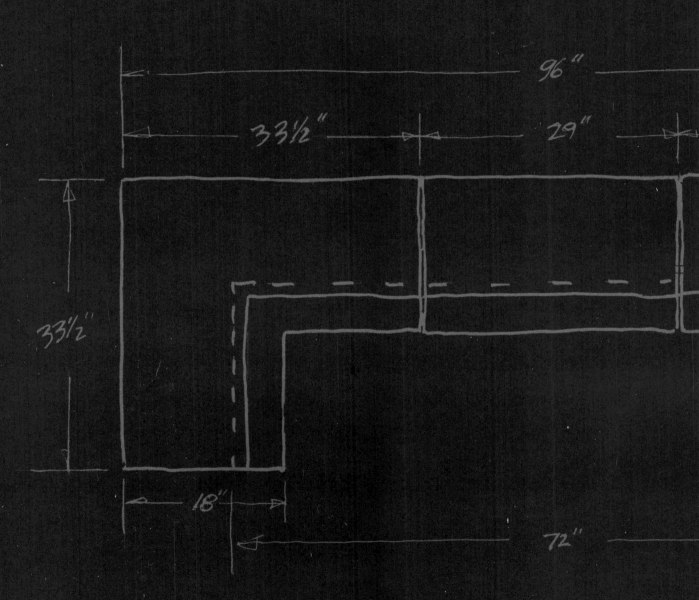

MAKING
CONCRETE COUNTERTOPS
WITH BUDDY RHODES
ADVANCED TECHNIQUES

Schiffer Publishing Ltd

4880 Lower Valley Road Atglen, Pennsylvania 19310

ACKNOWLEDGEMENTS

This book would not have even gotten off the ground without Tina Skinner and Doug Congdon-Martin. We wish to thank them for their creativity and focus, as well as our supportive colleagues at Buddy Rhodes Concrete Products: Heriberto Esquivel, Joshua Cruz, Matthew Mondini, June Mejias, Lorna Villa Seven, Mark Gunther, Steve Schatz, Keith Couch, and Jim Mason.

DESIGNED BY Laura Mikowychok. Type set in Zurich.

PHOTOGRAPHY BY Doug Congdon-Martin of Schiffer Publishing, Ltd., unless otherwise noted.

ISBN: 978-0-7643-3014-8
Printed in China

PUBLISHED BY Schiffer Publishing Ltd.
4880 Lower Valley Road
Atglen, PA 19310
Phone: (610) 593-1777; Fax: (610) 593-2002
E-mail: info@schifferbooks.com

For the largest selection of fine reference books on this and related subjects, please visit our web site at
www.schifferbooks.com

We are always looking for people to write books on new and related subjects. If you have an idea for a book please contact us at the above address.

This book may be purchased from the publisher. Include $3.95 for shipping. Please try your bookstore first.
You may write for a free catalog.

IN EUROPE, Schiffer books are distributed by
Bushwood Books
6 Marksbury Ave.
Kew Gardens
Surrey TW9 4JF England
Phone: 44 (0) 20 8392-8585; Fax: 44 (0) 20 8392-9876
E-mail: info@bushwoodbooks.co.uk
Website: www.bushwoodbooks.co.uk
Free postage in the U.K., Europe; air mail at cost.

INSIDE
THIS BOOK

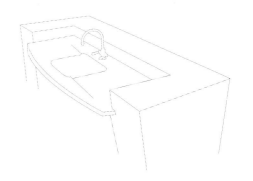

BUDDY RHODES
PREFACE

NOWADAYS, MORE OF US WANT our outdoor adventures to take place right in our own backyard, literally. In the last twenty years we have torn down dining room walls and moved dinner guests into our kitchens. We've invited them to join us for drinks and tasting while sharing the cooking experience. Forget that bar area in the living room with the martini glasses and the shaker; that's for black-and-white movies.Come on in while I'm cooking, and sit at the concrete counter for a chat.

Similarly, we now want to move everyone outdoors when the weather gets warm. However, we don't want to go back and forth into the kitchen and miss all the fun. No more dad at the barbecue and mom inside mixing up the potato salad; just move the kitchen outside.

Concrete turns out to be the perfect solution for the outdoor kitchen countertop. Hot summer sun tears up polished granite, warps other solid surfaces, and melts the glue in laminate. But concrete – now *there's* a material that's made for the outdoors. It may have come indoors recently, but you can march it right back out again.

This book was written with the goal of teaching more advanced techniques of concrete countertop fabrication, such as three-dimensional mold making, integral sink formation, curved edges, and built-in drain boards. An outdoor kitchen project meets the lifestyle of our new millennium, so professionals and advanced do-it-yourselfers alike will appreciate these step-by-step guidelines for new applications to incorporate into projects, even if they end up being put to use back inside for cooler days in the kitchen and bath. SA

DAVID SHEFF
FOREWORD

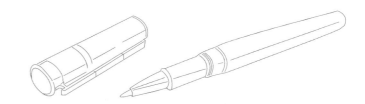

I GREW UP IN A TRACT HOME with pre-made Formica-clad countertops, plastic showers, and speckled linoleum on those floors that weren't covered in wall-to-wall shag carpeting. I had no idea that a beautiful home could change a life. I learned that it could when the California architect John Marsh Davis, renowned for his homes, wineries, and other buildings that evoke Frank Lloyd Wright and Green and Green, designed a home for my family. The home is filled with light, framed in board and batten, and set in a garden that is also designed by the architect. A critic called it "as perfectly constructed as a Chinese puzzle and nearly as intriguing." Living in it I have learned that a home can do far more than shelter us. It can bring with it inspiration and joy and a profound sense of peace. After almost two decades, I still stare at the details throughout the house and see new things.

The home is grounded by, and in fact defined by, tiled floors and concrete countertops that stop visitors in their tracks. They are designed and created by Buddy Rhodes. Before you enter the house you are met by Rhodes' concrete tiles in an entrance patio. The tiles are indescribably lush and warm colors that only Rhodes could have made. To say they are earth tones doesn't do justice to the breathtaking color. To say they are colorful doesn't do justice to the rich depth in each tile. I thought of them when I visited Tuscany and saw stunning villas that had been baked by the sun for centuries. The pattern continues inside through the front door in the entryway. Generally that's as far as I need to go before I feel a sort of thrill. A thrill from tiles? These aren't tiles. They are art. They, like the best art, tell a story and contain within them emotion. They stop me because they are exceedingly beautiful.

One of my favorite rooms in the house is all Buddy Rhodes, a sort of gallery of his work. His diamond-shaped tiles, grays and cream white, cover the floor. They form a pattern that appears three-dimensional. The Rhodes concrete countertop and open tiled shower that looks out onto bamboo are also gray and creams. The concrete invites nature in through the windows and it makes its own statement while respecting – somehow showing off – the only other materials in the room, oiled cherry and redwood.

The bathroom isn't the only room with Buddy's countertops. People who come over always marvel over the ones in the kitchen. They are built of smooth Buddy Rhodes concrete and are soothing and gorgeous. They are also remarkably functional. They and a good stove have turned me from an aficionado of take out to a decent chef. I love cooking on these countertops. A friend who is a food writer for Gourmet and author of a dozen cookbooks, who also has Buddy Rhodes countertops in her kitchen, told me that I'm not alone. "No one understands how a work surface can transform the experience of cooking," she told me. They provide a sort of center for the kitchen and ground the interior of our home in nature; concrete feels like the earth because it is made from it. Buddy's secret formula, however he pulls it off, has an effect that is both subtle and profound. At the same time it's heavy and light, striking and subtle. Over the years I have noticed how the countertops breathe, reflect and absorb light, and contain a richness and power that no other material has. They have only gotten better with age. DS

David Sheff is a journalist, veteran interviewer, and author of multiple non-fiction books on such varied issues of his time as video games and the Chinese technology and business revolution. He conducted the last major interview with John Lennon before his assassination, and has also interviewed Ansel Adams, George Lucas, Jack Nicholson, Steve Jobs, and Keith Herring, among many other celebrities, artists and visionaries. Sheff has written for Playboy, The New York Times Magazine, Rolling Stone, Wired, Outside, *and* Fortune.

A BRIEF
INTRODUCTION

I CAME TO CONCRETE as a potter, making vessels with clay. In art school, I made hollow clay bricks and formed tunnels and constructions with them. Requiring more and more bricks without enough hours in the day and night to do all the firing, I discovered the possibilities of concrete. In order to find a self-hardening, self-firing material, I set to work testing out mix designs, just as I had once designed my own clay formula. No more firing. Concrete bricks and blocks became my new vocabulary.

I liked the veins created when I pressed clay against the mold. With concrete, blocks could go bigger and be more complicated. Once out of art school, I began making furniture, planters, and hollow shapes, still using the method of pressing the material against a mold leaving voids and texture.

In the mid-eighties I turned my attention to a new idea: concrete countertops. If I filled in the voids and pits, sealing and polishing the finished surface, I could create buttery smooth countertops with many combinations of color and visual texture. Like pottery to hold in the hand, I found the warmth of clay surfaces using concrete finishes. Soon I added the alternate hard-trowel surface, still showing the hand of the artist and the warmth of the crafted material, without the veins.

After twenty years of making concrete countertops and furniture, tiles and planters, I had formulated a concrete mix that was ideal for my methods. With a desire to make it easier to produce the multiples with the reliability I required, along with an intention to let go and pass on my techniques to others, I bagged my mix and developed a line of products. Our staff began to teach workshops, sharing methods and materials to help others find success and avoid the many mistakes I made in my early endeavors.

In this book about advanced techniques, I'm going back to the core concept of my work: hollow blocks. You don't have to stay with flat surfaces. The methods and the mix formula used here are developed for molding and crafting by hand. The process is still in many ways quite simple, with focus on the material rather than an engineered mold. I want you to think about concrete in a new way. Work it. Get your hands in it and don't just go flat; go hollow! You can go up a vertical surface; imagine the possibilities. You can go crazy: running walls, built in seating areas around a garden. There's no gravel in the mix, so it's more like mortar. That's one of the reasons you can work it up a vertical surface or make an integral sink wall. If you're not going vertical though, feel free to add recycled aggregate of all kinds, or pea gravel if you're casting in place; I use about 20 lbs. per bag.

With the goal of demonstrating my enthusiasm for a new way of looking and thinking about this friendly material, I have chosen a freestanding outdoor kitchen. The structure of this outdoor kitchen is a series of hollow blocks shaped to hold a countertop. You will learn some more advanced techniques not shown in our first countertop book: three dimensional mold-making, the vertical use of BR Concrete Mix, a simple method for drawing a curved edge on a form, and how to add dimension to a concrete slab through an integral sink and drain board. Remember, we're "packing" not "pouring." We don't have to worry about additives or suspending reinforcement within the mold. Stir up some Mix with water and color and pack it into molds of 3/4" melamine screwed together – advanced "play-dough" techniques.

We will start out with an idea, in this case an outdoor kitchen, and, after drawing the dimensions out on melamine with a pencil, we'll make our project together. We'll play with texture and go rustic with open voids and pits where there will be no food preparation. We do that often with garden furniture, and, if we're lucky, we might even get some moss to grow in the nooks and crannies. Imagine your project; think of building a box that is the outside of what you want to make. The air inside the box is what you're making. You'll pack the inside of the box with concrete and reinforcement, focusing on working the material. When you take off the mold you'll have your structure. Think hollowness.

If you don't want veins on the structure, you can paste over the vertical surfaces with a slurry coat of concrete after the mold comes off.

If you want a hard trowel or lightly ground countertop instead of a veined one, cast the counter with those methods detailed in *Making Concrete Countertops with Buddy Rhodes*.

You can take off with these ideas and fly on your own creative current. Flexibility of form and color is the great advantage our medium. You are the artisan who will make the functional piece for the home, with love and care for the family who will enjoy it for years to come. BR

WHAT YOU'LL NEED
TOOLS & MATERIALS

I HAVE LISTED the tools and materials I used to plan, form, cast, and construct the outdoor kitchen project. Ideally you will have these tools and materials on hand to construct your own project based on an inspired take off from this one. However, improvisation is often the key to success, and sometimes a substitute can be found for an expensive tool that doesn't live in your workshop.

The most important thing is that you be prepared in advance. Think through your process and buy, borrow or improvise the tools you'll need before you've mixed your concrete, or before you are in the midst of applying your finish.

No, the most important thing is following the manufacturer's directions. Well, actually, the very most important thing is heeding the manufacturers' warnings, and making sure you have proper safety equipment and gear.

MATERIALS

Graph paper, ruler, and drawing pencil
Architectural circle template
T-square
Level casting tables
Plywood
3/4" thick 4' x 8' two-sided sheets of melamine
Bondo
Newspaper
2" screws
Vinyl tape
4" thick Styrofoam
1-1/2" thick Styrofoam
Polyester resin
Lacquer thinner
Putty knife
Small nails
100-grit sand paper
Rubber drain knockout
Galvanized Ladur wire reinforcement
Polypropylene DuraFibre
Latex gloves
Leather gloves
Galvanized wire mesh reinforcement
Dust mask and safety glasses
Measuring bucket to measure water
    for every bag used

Plastic drop cloth
Mud tray
Knee pads
Polypropylene reinforcement fiber
Soft Cotton rags
Polypropylene fibers
Wood scraps
Water sprayer bottle
Mud tray (as used for sheet rock)
Raingear for wet grinding
Scotchbrite pad
Hand sand pads, 400 grit
Lacquer sealer
Small facial respirator

TOOLS

Table saw
Band saw
Circular saw
Cordless screw gun
Jigsaw
Drill press
Palm sander
Chop saw
Wire cutter
Tape measure
6" putty knife
Clamps
Wood float
Pry bar
Brick stone
Potter's kidney-shaped tool for smoothing
    curved areas
Variable speed wet polisher with 400-grit
    diamond pad
Diamond cup wheel on wet grinder.

BUDDY RHODES PRODUCTS

Fiberglass sink mold
Buddy Rhodes Concrete Mix
Color
Acrylic Additive
Color Paste
Penetrating Sealer

SUPPLIES

SPECIAL WORKTABLES

I make my own worktables from plywood and 2x4s. Not to be confused with casting tables, these worktables are an alternative to sawhorses, are always level, and withstand being cut into and across. They are movable, stackable, and used temporarily to provide much more support for pieces of plywood than sawhorses. I can cut a piece of melamine on two or four of them and cut right through quickly and easily. When I don't need them I just stack them up and set them aside in the shop.

To make three handy worktables, start with one 4' x 8' section of 1/2" plywood and six 8' 2x4s. Rip the plywood into three sheets, each 15-7/8" wide, and then cut into 23-1/2" lengths to make 12 panels. Cut the 2x4s into 31-3/4" lengths to form the legs. Screw the assemblies together and voila!

A F T E R   T H E   I D E A :
THE PLAN

MOVING FROM GRAPH PAPER TO REALITY is a big step. When I started to make the molds, I realized the countertop heights needed adjustment. I knew I wanted one for counter area for preparing and cooking and one for a bar top and socializing area. Not a cook myself, I went next door to Moshi Moshi, the bar and Japanese restaurant next to my shop, and measured the surface heights. I measured the bar heights for sitting on stools and the height for food preparation. Then I adjusted my planning for the molds. You are seeing here the second phase plans.

When planning a three dimensional mold with a drawing, you have to think upside down and backwards. Think of making the void inside the form. You are encapsulating the shape you want. Make the drawing of the shape you want, add on the thickness of the melamine, and that's the drawing outline. The more you do it, the easier it gets. Draw your project on paper as best you can, then if more ideas and widgets come to mind that you didn't think about in your drawing, adjust for them during mold-making.

My idea was to have an interlocking system, like dry laying stones. Nestled together, I wanted it to be able to move a bit with the nature around it. I didn't want to grout or glue the pieces, taking advantage of a certain freedom of being outside, and in the case of my freestanding unit, not being attached to a wall. Concrete doesn't rot and withstands the weather famously well, although in northern climates you can throw a cover over for extra protection from sitting water and freeze-thaw cycles.

I planned a free-standing unit with an integral sink. We haven't talked about plumbing, but you'll have to include that in your own planning. My thinking was to wheel up a propane barbecue unit like the one I have at home, and to build another work surface to match on the other side of the barbecue. Alternatively, that other surface could have a drop-in stove; there are amazing options on the market now. How about a wood-fired pizza oven in your outdoor kitchen plan?

The plan here is meant to be inspiration. Remember, you can cope with the surprises that come up as you go. For example, I hadn't planned for the plumbing gullet. I realized I needed it as the project unfolded, so I adapted for that after the drawing phase by placing a piece of foam for the gullet in the mold. Let the requirements of your site determine your design.

Because I often make changes as I go along in the mold process, I always choose to pre-cast. I can re-cast a piece; if I make a mistake in casting. All the advantages of your workshop will be at your disposal if you can build the pieces in the shop or garage and carry them to the site. Yet, a rather massive project like this can't be moved through the house into the backyard very easily. Alternatively, consider taking the molds, with unscrewed pieces if need be, and cast them where they will be installed when dry. They will then fit together like a puzzle if you have cast them according to your plan.

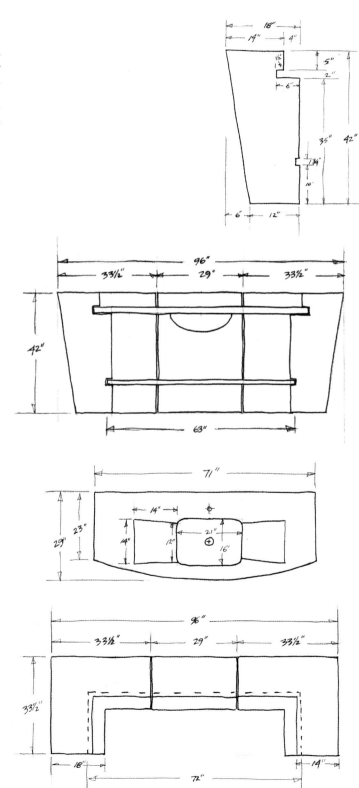

BUILDING A VERTICAL MOLD

THE CENTER UNIT

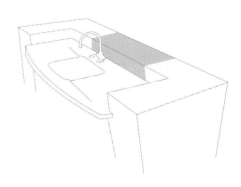

IN THIS SECTION we will form and build the melamine mold for the smaller center unit. We will draw an outline of the side profile, front, and back pieces with pencil, and proceed to cut out pattern pieces based on our design plan. Step-by-step, we will put together our center unit mold, providing the outside form for our keystone piece.

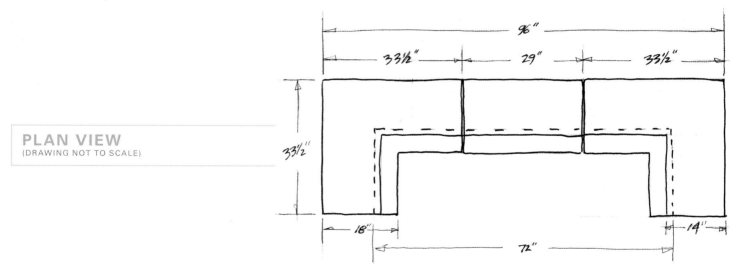

PLAN VIEW
(DRAWING NOT TO SCALE)

We do the sides of the smaller, center mold first. This section right here is our profile. I'll cut out two of these first. The profile pieces of the center unit are measured and drawn onto a piece of 4' x 8' foot melamine with pencil.

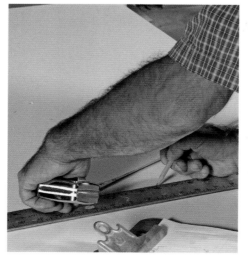

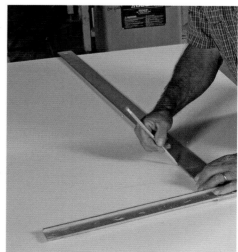

The straight lines of the mold are cut on the table saw.

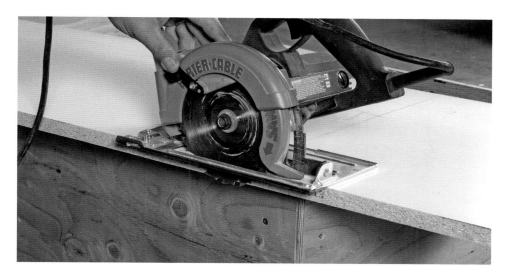

The angled cuts of the mold are then cut out with a hand-held circular saw. The saw is set so that it cuts just slightly deeper than the 3/4" melamine.

Cut out the corner notch that will form the ledge to support the sink counter cut, using the circular saw. Cut just to the line.

A handsaw gets the final cut in the corner.

On melamine, plot out the front and back pieces of the unit mold using pencil, tape measure, and T-square.

The pieces for the front and back are all the same width, so a melamine board is cut to size on the table saw.

Use the circular saw to cut out the profiles of the front and back mold pieces of the center unit.

Using a strip of melamine board, mark where the overlap will be when it is attached to the end piece when the mold is assembled.

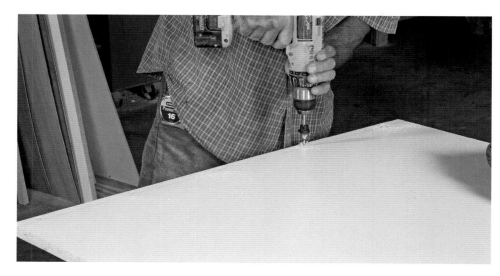

Use a countersink drill bit to pre-drill holes along the sides about six inches apart.

The back side of the center unit mold is assembled first.

The center unit is turned over, and the front is test fitted.

The front was cut too long on purpose in order to compensate for the angle needed on the cross cut. After a test fit, the overhang is marked with a pencil...

...and cut to fit, adjusting the angle of the saw blade to get the beveled cut. The front is screwed on to the assembly and we flip the piece over and get to work on the back again.

Carefully measure for the shelf pieces.

A piece of melamine is cut to fit the shelf face.

The screw holes are marked with a pencil …

… and pre-drilled on a drill press.

Measurements are made for the shelf. Again, a piece is cut to fit, the drill holes marked and countersunk before assembly.

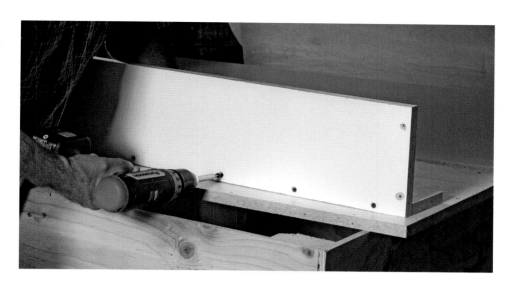

Lay the shelf piece on its side and stand the face piece along one edge. Attach the face piece to the shelf piece, screwing through the face piece.

Put the shelf assembly into place. Pre-drill and countersink screw holes …

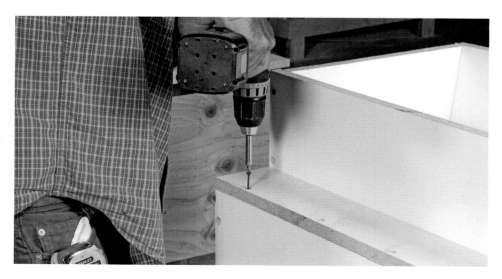

… and attach to the mold.

The top piece is measured, cut, and attached to all but the front panel, which will be removed for the pour.

Square the mold using a carpenter's square for accuracy.

Turn the mold upside-down. Bottom up, we only need a frame around the base for the 2" wall thickness at the base.

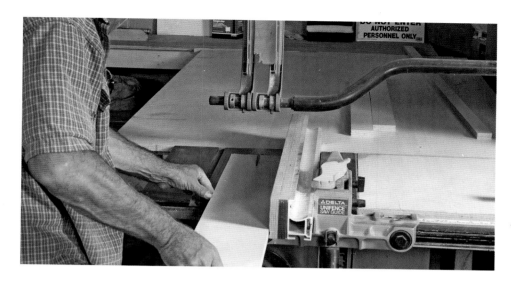

Because we want a 2" lip on the bottom of the mold, we cut melamine strips that are 2-3/4" inches wide to allow for the thickness of the melamine mold walls.

The 2-3/4" inch melamine strips are cut to length and attached to the base of the mold, via pre-drilled, countersunk holes.

The mold for the center unit in the sink surround is almost complete. The front piece will be removed for the packing process and then reattached while wet.

FILLING COUNTERSINK HOLES

FILLING COUNTERSINK HOLES and exposed board ends with Bondo prevents the concrete from sticking to the raw molding and prevents unsightly seams. On these pages I have illustrated Bondo application techniques, which we will then apply to our mold.

Mix Bondo body filler with cream hardener.

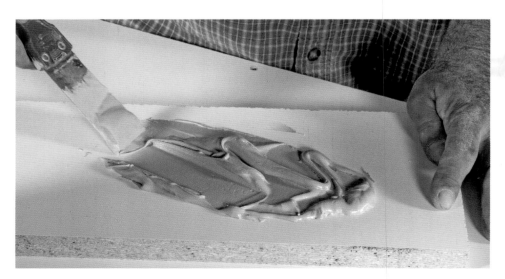

Use a 2″ putty knife to integrate the ingredients.

Apply the Bondo mix to exposed edges of melamine board and fill the countersink holes on the insides of the mold.

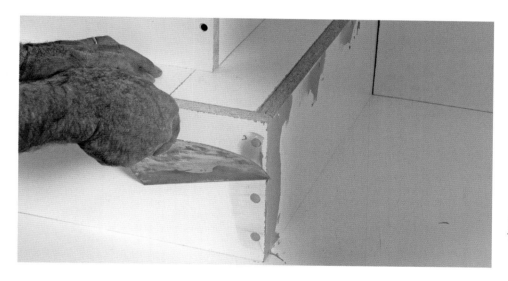

Use a 6" putty knife to achieve a smooth finish.

After curing and sanding, the joint is ready for packing with concrete.

Fill all of the countersink holes and exposed ends on the center unit mold you've created. After curing for one hour, sand the Bondo surfaces smooth with a palm sander.

Cut two strips of 3/4" plywood, ripped 4" wide, at a 45° angle and attached to the center mold to provide support for the vertical casting. Be sure to attach these supports with screws that will not penetrate through to the inside of the mold.

COUNTER & LOWER SHELF KNOCKOUT FOR THE CENTER UNIT

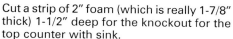

NOW WE'LL PLACE the knockouts for the notch in the center unit. This will hold the counter and lower shelf. The sink knockout used 2" Styrofoam, which has an actual thickness of 1-7/8". The tape added to the foam knockouts made it a little bit thicker. The sink counter will be 1-3/4" thick, which will give it some room for sliding it into place. The sink counter is 71-1/2" inches long.

Cut a strip of 2" foam (which is really 1-7/8" thick) 1-1/2" deep for the knockout for the top counter with sink.

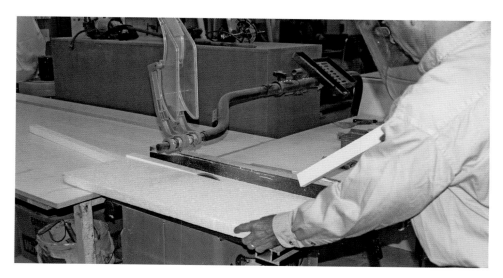

Cutting foam on a table saw is more dangerous than cutting wood. It will fling out and draw your hand into the blade, so be very careful.

The counter knockout strip is measured to fit and attached to the mold with screws.

Cut two spacers from melamine or wood to mark the distance from the top knockout (now on the bottom of the upside down mold) and the knockout position for the bottom shelf.

The lower shelf knockout was created from 1-1/2" foam, which will form the depth of the slot. It was cut at 1-5/8" for the height. Attach the lower shelf knockout foam to each side using the spacers as a guide.

Screw the knockout in at equidistant points, again using the spacers to ensure the proper level for the shelf bracket.

Concrete will reflect the rough surface of foam so we use vinyl tape to cover the foam. Vinyl tape creates a smooth surface and leaves a nice finish. Most concrete supply houses carry it.

PREPARING ROOM FOR PLUMBING

FOR THIS MOLD we decided to create a channel that would extend from below the sink straight through the ground, allowing plumbing to be run underground to an outdoor sink, or from the basement if this unit were brought indoors.

Cut a 6" wide gullet from 4" foam to span the distance from the counter shelf to the base. The foam is cut using a handsaw rather than the table saw because of the thickness. A square ensures a straight cut.

Tape the foam blocks.

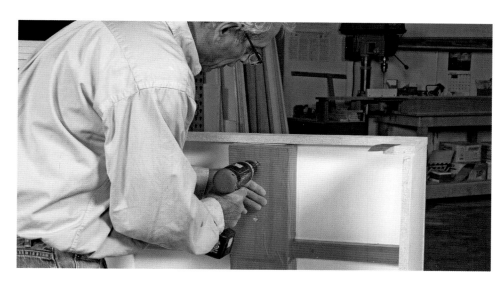

Attach the gullet blocks to the mold with screws.

Tape over the screw holes.

The center unit mold is complete.

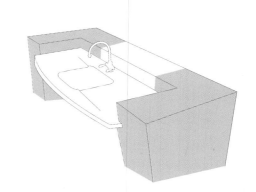

THE END UNIT

IN THIS SECTION we will build one L-shaped mold to be cast twice for the two corner pieces.

LEFT SIDE VIEW
(DRAWING NOT TO SCALE)

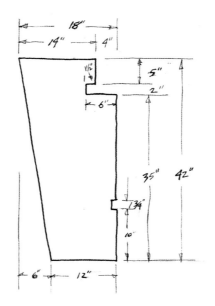

For the second mold, the end piece, we start with the same process, measuring and cutting the side pieces. These pieces are identical to the sidepieces of the center unit. In the case of the two end units, we build one mold and use it twice. Melamine molds are good for only two castings.

Layout the inside corner panels of the end unit mold, and cut them out.

Connect the two inside corner panels using countersunk screws.

Connect the side panels to the center.

Because the front panels are angled, cut the pieces longer than anticipated and then cut down to fit.

The circular saw is used to cut out the outside panels to length.

Pre-drill and countersink the screw holes.

Using a board to support the assembly, tack the front panels into place. They should overhang at the top and bottom.

Screw the panels in place.

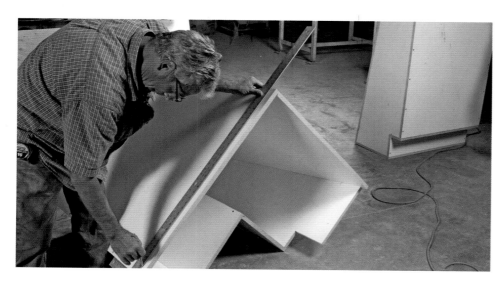

Mark the overhang of the front panel using a T-square.

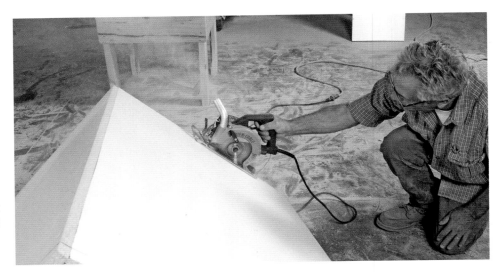

With the blade set at an angle to give a flat cut, remove the overhanging portion using the circular saw.

The same process is repeated on the bottom of the mold.

The shelf portion of the mold is measured, cut, and assembled, then …

… attached to the mold.

Measure and cut the top.

Countersink screw holes along the edges of the mold's top.

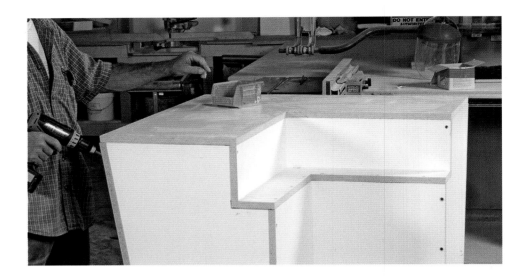

Attach the top of the mold.

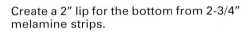

Create a 2" lip for the bottom from 2-3/4" melamine strips.

A smaller piece is added to the front for easier attachment and removal of the front panel of the mold during the process of packing concrete. Again, the front panel will be removed for the pour and then reattached.

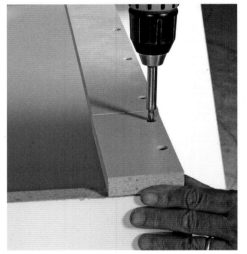

Attach the final strips to the base of the mold.

The mold construction is nearly complete. Remove the front panel and clean the mold.

A close-up of the angled shelf mold. We'll need to fill the countersink holes and exposed board edges with Bondo on these molds, just as we did for the center unit (see page 20-21 for Bondo application techniques). Remember to allow the Bondo to cure for one hour before sanding.

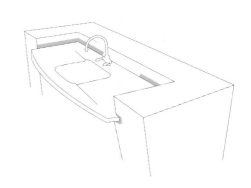

MAKING SHELF KNOCK-OUTS ON THE END UNITS

The same process of adding Styrofoam knockouts is used on the center mold (see pages 22-23). Use the same spacer bars to ensure accuracy!

Do a final dry fit of the removable front panel for the end unit.

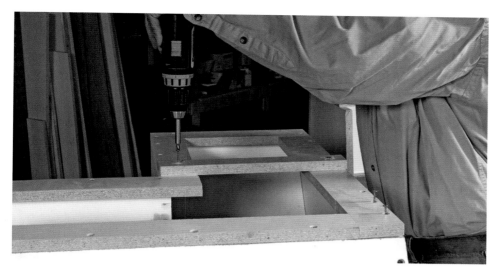

The weight of the concrete may distort the mold, so reinforce the top to hold the form. Screw a crossbar to the bottom of the end unit mold to ensure its stability and help it keep its shape for packing.

Mark the front panel at 3/4" along the edges to act as a press-to guideline during the packing process. You don't want concrete to be in the way for the re-attachment. A straight edge and pencil provide the pour line.

ADVANCED
BUILDING MOLDS WITH INTEGRAL ELEMENTS

PLOTTING THE COUNTER
AND RADIUS EDGE

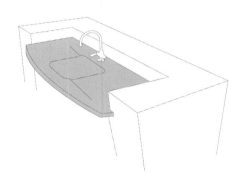

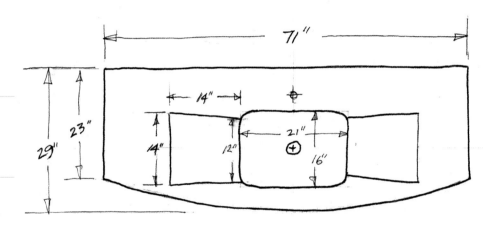

Use a straightedge to draw a line 15 inches from the edge of a 4′ x 8′ sheet of melamine, allowing the counter to have plenty of room on all sides.

Find the center of the 71-1/2"-long countertop and mark.

Draw the centerline.

To create the curved front, you need a long compass, and you need to continue the centerline back far enough to get a radius of about 12 feet. A second worktable will allow you to extend the line.

Use a long straight edge and a pencil to mark the centerline's continuation.

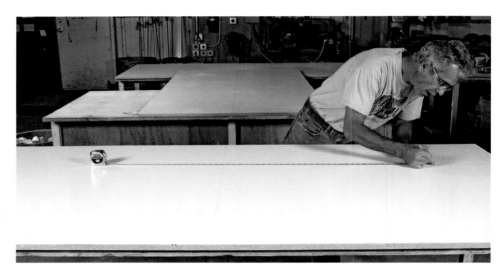

Figure out the width of the countertop and measure from the centerline to each side.

Mark the edges. Note that the countertop length is a half inch shorter than the shelf slot allotted in the base so that it will slide in to its slot. Make a 1/4" allowance on each side.

Using your T-square, plot and draw the sides of the radius countertop.

Our compass is set to 10' 9". We calculated this on graph paper beforehand, and then adjusted it. While I was actually making the mold, I tried 8', and then went to 10' and wound up with 10' 9" by looking at it and finding a curve that looked right for me. The drawing that you see reflects the final decision.

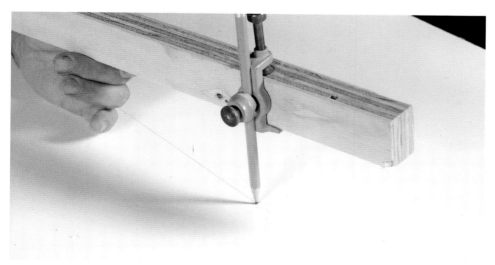

A close-up shows the writing end of the compass, created using a standard clamp available in hardware stores.

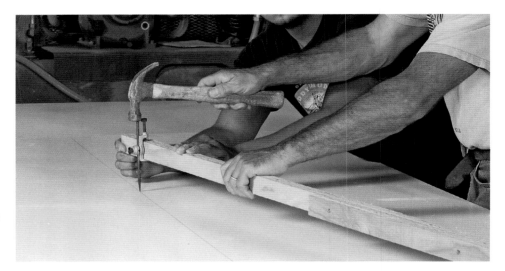

Tap the far end of the compass lightly into the board below to anchor it.

PLAN VIEW WITH COMPASS FOR RADIUS EDGE

(DRAWING NOT TO SCALE)

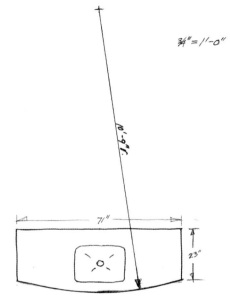

$\frac{3}{4}" = 1'-0"$

10'-9"

71"

23"

Draw the radius edge carefully. It helps to have someone holding the pivot in place.

PLOTTING THE INTEGRAL SINK AND DRAINBOARD

We will be using a fiberglass sink mold manufactured for Buddy Rhodes Studio.

Carefully measure the center of the sink mold. Here the bottom of the fiberglass mold is shown. Because it's flat, the measuring starts here.

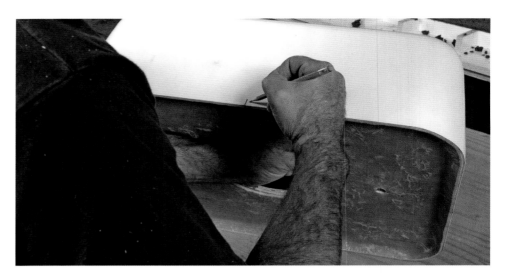

Use a straight edge to carry the centerlines to the sides and to the front. Mark the centerlines on all four sides.

Measure 5" back from the front edge of the countertop and mark the position of the front edge of the sink at the centerline.

Center the sink mold on the melamine board and trace around it.

At the sides, locate the centerline on the sink mold and mark it on the melamine.

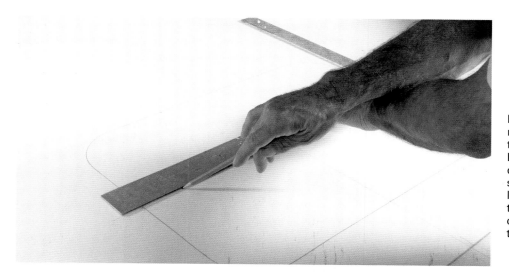

Extend the lateral centerline across the melamine. Measure 14" from the side of the sink and mark it on the lateral centerline. This will be the far side of the integral drainboard. Use a T-square to create a straight line, intersecting the lateral centerline at 14" from the sink edge. Mark along this line at 8" from the lateral centerline, creating the 16" flange at the far side of the integral drainboard.

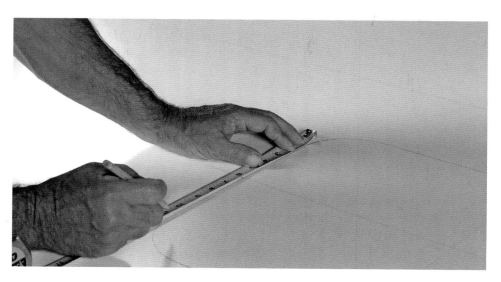

Along the sink edge, measure 6" from center and mark at each end, creating the 12" width of the integral drainboard beside the sink.

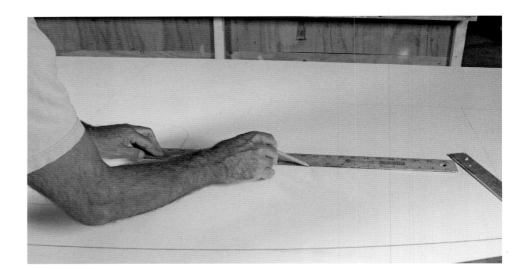

Now connect the dots.

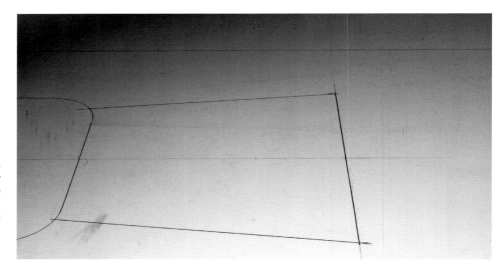

The flared shape of the integral drain-board and the outline of the sink are now penciled in. The darker lines represent the drainboard area, which will be cut out. Repeat this process on the opposite side of the sink.

Drill a hole just inside the line of the sink where it intersects with the drainboard, in order to insert the jigsaw. Do the same at the other corner and at one of the outer corners of the drainboard, staying inside the outline of the drainboard.

Use a jigsaw to cut out the drainboard area. The drainboard should lift out like this. Mark the first drainboard area "left" or "right" so you'll know where it goes later.

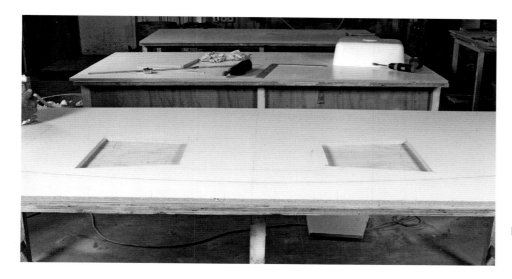

Remove the two drainboard areas.

Place newspaper under the holes to prevent attaching the mold to the table with resin.

Place a length of wood 1/2" thick against the sink profile to act as a wedge.

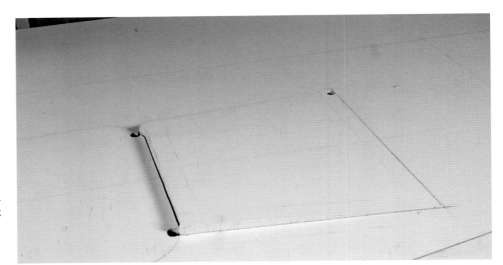

That 1/2-length of wood creates the drain-board slope when the cutout is placed back in its hole.

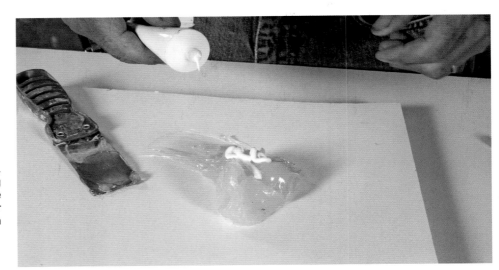

Mix a polyester resin (Akemi, Touchstone, Tenax) with catalyst. You want something that has body, and won't fall into the cracks. This bonding resin is used for marine applications and can be found in hardware stores.

After mixing, use the resin to butter the sides of the drainboard hole and the half-inch wedge board.

Place the drainboard mold and remove the excess resin. Repeat the same on the opposite side.

Fill in the cracks and the pre-drilled holes with resin. Then wait about half an hour for the resin to dry. Lacquer thinner can be used to remove excess resin. It will remove the pencil marks, so you want to be careful about that.

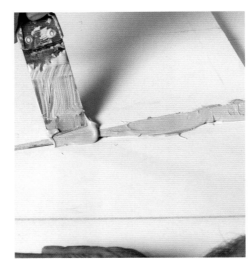 

Apply Bondo with a 2″ putty knife to create the angles at the sides of the integral drainboard.

Use a larger putty knife to finish and smooth. Allow the first application to cure for about an hour.

Carve away the excess Bondo using a putty knife.

Sand the first application of Bondo with a coarse sandpaper on a block.

Apply a second coat of Bondo to fill voids and perfect the angles.

Use a palm sander to smooth the drainboard mold.

Spray the Bondo areas of the mold with a shellac-base white primer. By making it the same color as the melamine, it allows you to see the deviations and bumps. And it looks nice.

MAKING THE MOLD EDGES

Because the countertop will be 1-3/4" thick, rip 1-3/4" mold walls from 3/4" melamine on the table saw.

Countersink screw holes into the edges of the mold walls. We will use 2-1/2" screws so that they will go through the 1-3/4" mold edge and into the base of the mold, but not through.

A 10' steel straight-edge helps align the spine, or back wall of the mold.

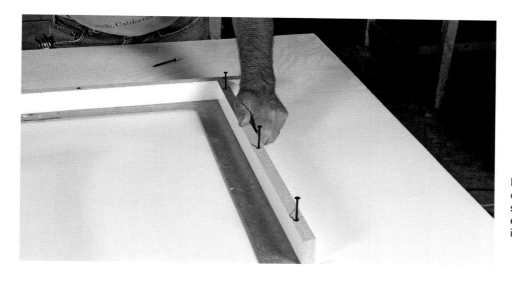

Measure the side walls and cut. On the chop saw bevel the front edge of the sidewall to match the angle of the curved edge. Screw the sidewall of the mold into place.

Cut out blocks from 3/4" melamine, 1-3/4" wide. These blocks will support the curved wall in the mold.

Countersink screw holesin these bracing blocks for the radius edge.

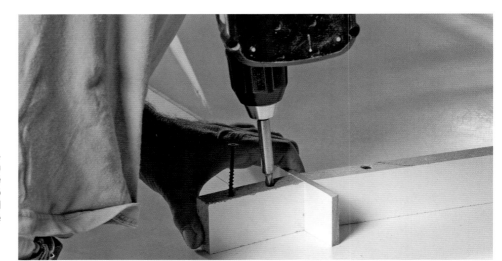

The curved wall is made of 1/4" melamine, ripped to 1-3/4". Cut a strip long enough to form the curved wall, allowing a couple of excess inches on each side. Affix two blocks, one to each of the beveled sidewall molds. Bend the 1/4" melamine strip to the radius line drawn earlier.

Fix one block to the centerline. (Note that the radius edge pencil line should show on the inside edge of the 1/4" melamine strip).

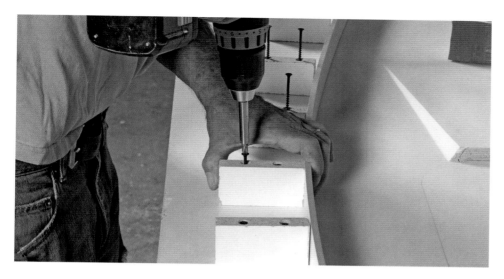

Now affix the rest of the blocks, evenly spaced and placed so that the melamine strip can be attached just outside the drawn radius edge line.

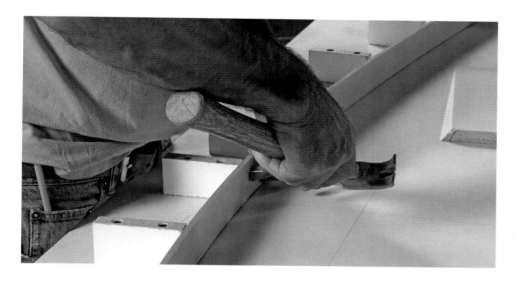

Use finishing nails to affix the 1/4" melamine strip to the blocks behind it. The nails are so small that we won't need to use Bondo.

Fit the sink mold into placer.

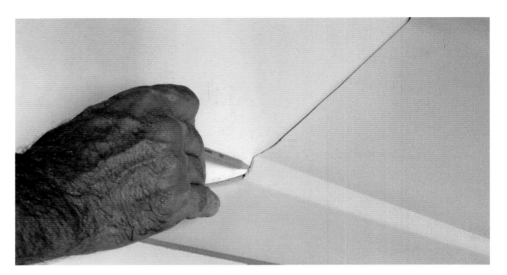

Use a pencil to mark places where the sink doesn't fit properly against the raised drainboard.

A little bit of dried resin and wood needs to be carved away with a knife where the fit is too tight.

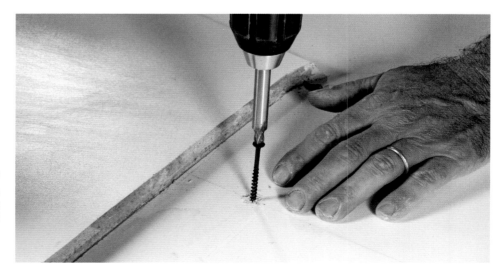

Drive two screws through the base of the mold on either side of the centerline and remove them. The holes will work as a guide when securing the sink mold from beneath.

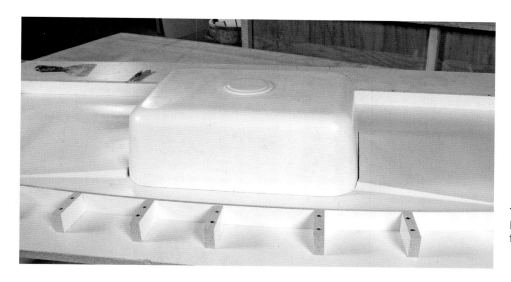

Turn the mold upside-down on the worktable, so that the sink mold can be accessed from underneath.

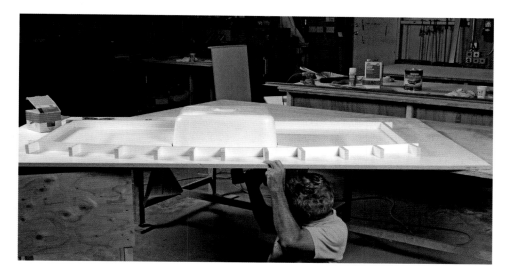

Screw the sink mold to the base.

Use 100-grit sand paper to smooth the mold. Take time to really go over it and make sure it's perfect.

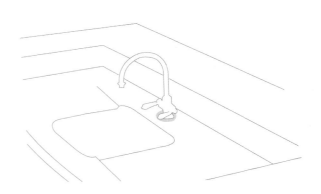

CREATING A FAUCET KNOCKOUT

WE NEED TO MAKE holes for our plumbing. Once you know which faucet you will be using, you are ready to proceed with making the displacement for the holes. Although you can drill them later, it's preferable to cast them into your mold.

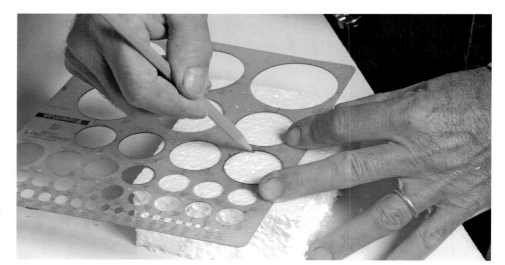

Using an architectural circle template, trace a 1-1/2" circle on 1-7/8" foam.

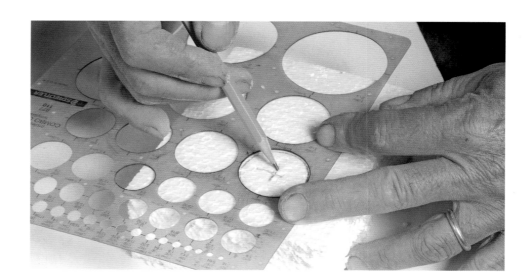

Mark the center.

Cut out a cylinder using a bandsaw.

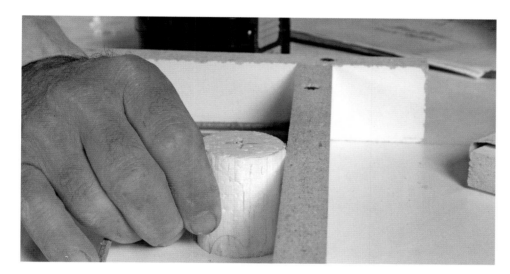

Hold the cylinder next to the mold wall to test for height. This foam piece is a little too tall.

Use course sand paper to reduce the height of the cylinder until it is flush with the mold wall.

Use the corner of the mold to ensure a level, perfect finish when sanding.

Wrap the foam cylinder with vinyl tape.

Trim the tape with a knife.

Measure and mark the knockout position on the mold.

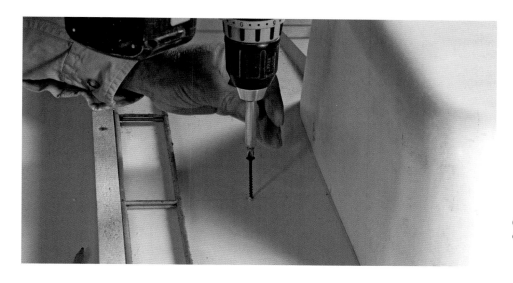

Create a starter hole in the melamine base of the mold using a screw

Drive a screw through the center of the knockout cylinder.

Use the starter hole to align the screw and affix the knockout.

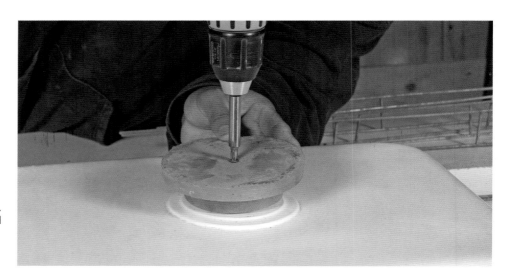

Affix drain knockout to center of sink mold. We are using a rubber mold manufactured by Buddy Rhodes Studio.

MAKING A SPECIAL SCREED TOOL

SPECIAL ATTENTION IS NEEDED with the back edge of this sink, because the base behind is angled and clearance is limited. You don't want the sink to hit the base unit. Therefore, we'll create a special screeding tool to ensure that the concrete sink is only 1-1/2" thick.

You'll need three strips of plywood. A small strip will act as a stop for the outside of the mold. Attach this first.

The two larger strips of plywood should be attached with a single screw, as shown. Using the faucet knockout and mold wall as a level for the base of your screed, measure to get an angle that will create a 1 ½-inch clearance for screeding.

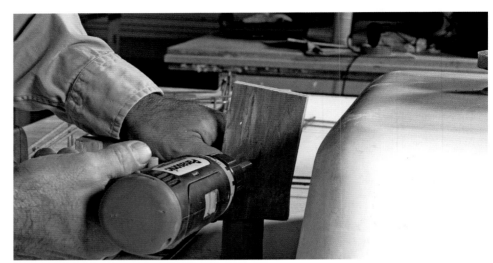

Mark the angle with a pencil, and add a second screw to secure the two wood strips at the proper angle.

Remove the protruding edge with your handsaw, and sand it smooth.

The finished screed tool.

THE LOWER SHELF

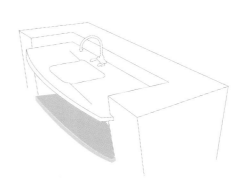

THE BOTTOM SHELF will be just under 1-1/2" thick. The thickness is a result of the 1-5/8" thick foam used for the knockout. It's 63-3/8" long.

After measuring back from the leading edge of a melamine board, use a straight edge and screw down the first 1-1/2" strip of melamine for the front of the bottom shelf.

Attach the first side wall of shelf mold.

Attach the second side wall of shelf mold.

Attach the back wall of the shelf mold.

REINFORCEMENT

WE WILL PREPARE REINFORCEMENT for the flat shelf and the counter with integral sink and drain board in advance. However, we'll measure and cut to fit our vertical molds after the initial layer of concrete has been placed. Because the measurements are so variable based on the thickness of the concrete and the many angles, it is easier to do it this way.

You'll find more information about reinforcement in the section on Mixing and Packing. For basic reinforcement theory and techniques, you can refer back to *Making Concrete Countertops with Buddy Rhodes.*

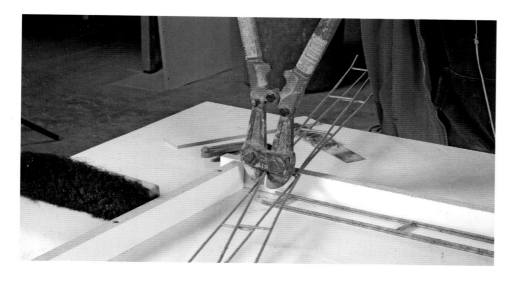

Cut lengths of galvanized Ladur™ wire to fit the flat areas of the counter and shelf molds, leaving about an inch at each side. You could also use threaded rod, or 9-guage welded wire mesh.

Bend and fit Ladur™ wire to reinforce around the sink mold.

You'll want to figure on about ¾ of an inch of clearance from the mold. Create four pieces like this, two on each side.

BELOW LEFT Clip rungs and side of Ladur™ wire to allow it to curve while lying flat and reinforcing the radius edge.

BELOW RIGHT Double the reinforcement with a short length of Ladur™ wire on top.

Place expanded wire mesh (a diamond wire or stucco wire) over the drain knock-out and use a marker to trace the outline of the hole to be cut.

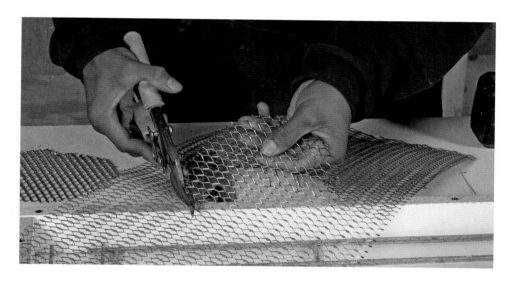

Use wire cutters to make a hole in the wire mesh.

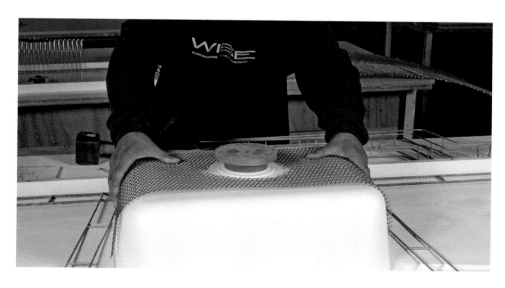

Bend the wire mesh to fit.

Two such layers are cut for strength.

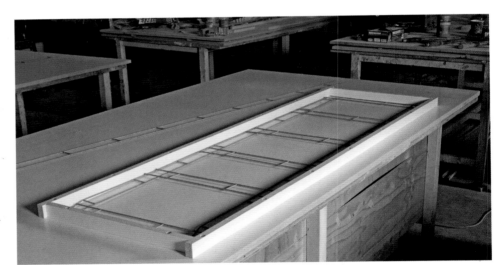

Cut Ladur™ wire reinforcement for the bottom shelf, too.

MIXING & PACKING

PROTIP

MAKE SURE your mold is as beautiful and perfect as it can be. Blow out or clean the mold before final casting to eliminate all dust. Have all your tools on hand, and all your reinforcement cut and ready.

TOOLS TO HAVE ON HAND FOR CASTING:

SUPPLIES

Mixer	6 inch putty knife
19-gallon large tub with handles (common party tub container; aka Muck basket)	Dust mask
	Measuring bucket to measure water for every bag that we make
Latex gloves	Plastic drop cloth
Leather gloves	Knee pads
Diamond wire mesh	Polypropylene rein-
Ladur™ wire	forcement fiber
Wire cutter	Diamond wire mesh
Tape measure	Clamps

MIXING CONCRETE FOR THE CENTER UNIT MOLD

THE BUDDY RHODES SYSTEM calls for one jar of color to two bags of concrete, plus four quarts of water per bag of mix. For hand packing, we need our concrete in stages. So we don't want to mix too much at one time. Hand mixing is the best way to integrate the ingredients when blending small batches, two bags at a time.

Add the color to the measuring bucket first.

Use a hose to clean the jar while filling the bucket. We need almost to two gallons.

Dump the water and color into the mixing bucket. A final rinse of the mixing bucket should bring the water content to two gallons.

Add the first bag of mix, and mix until well combined.

Add the second bag and mix to a clay-like consistency.

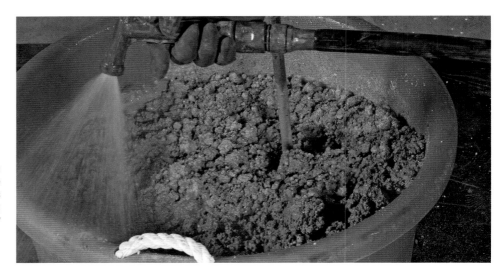

Depending on weather conditions, you may need to add a little bit of water, but be careful. You want a very dry blend. It has to be just right. If it's too wet its just going to fall down. It won't stick. This is also how you get the veining.

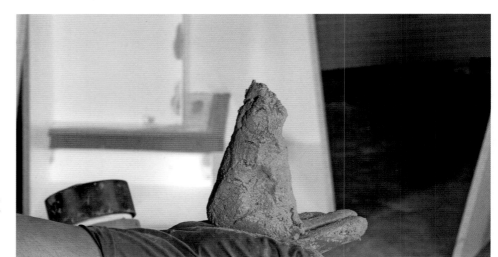

A cone demonstrates the zero slump property of the mix when properly blended for vertical application.

HAND PACKING

THE HAND-PACKING METHOD is demonstrated on Plexiglas to show the desired effect. It's important to press the pieces tight. The goal is to have voids, but not too large. The tendency is to make balls of the wet concrete, but this creates "elephant turd" syndrome, or a biscuit-like effect. The end result is a look that consists of too many round shapes, with large voids. It's important to start with small, irregular shapes and to press them tight. But not too tight, or you won't have voids! Practicing on Plexiglas is a good way to develop a feel for a technique that is executed blind, the results of which usually not revealed until the next day.

The process of hand packing is demonstrated on Plexiglas.

The wet pack shown from the other side of the Plexiglas.

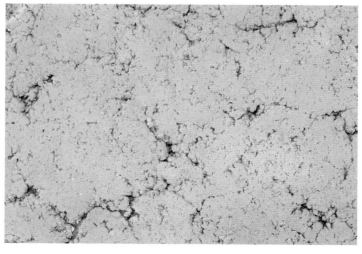

Dried and removed, a close-up of the piece cast on Plexiglas™.

PACKING THE CENTER UNIT

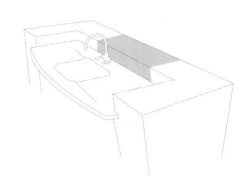

REMEMBER, we're packing this mold upside down. The mold is built up from the bottom. Kneepads are very helpful to have. We didn't put form release on because we didn't want it slipping. Also, we didn't pre-cut reinforcement wire since the interior measurements will be variable after the first layer of concrete is added. We will follow the same method for the center and corner vertical units.

Starting in what will be a top corner, above the counter slot, begin packing from the bottom up. You want a crisp corner, so make sure it's packed tightly in.

With the bottom filled, move above the shelf knockout, starting again in the corner.

Fill the vertical walls moving up.

Pack the top corner of what will form the base of the unit.

When the vertical walls are covered, go back over the corners, using your knuckles to pack them tight. Work from the base to the top.

Using a closed fist, pack the flat surfaces, bottom to top.

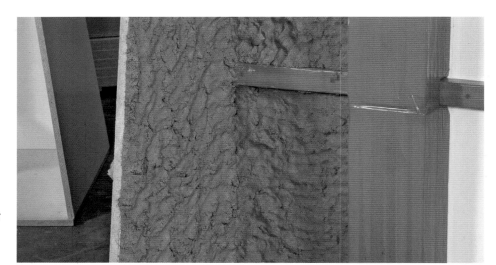

Go back and, beginning at the base, cover the gullet and shelf knockouts.

Pack concrete tightly under the top shelf edge that will form the base of the unit.

Packing of the front panel should begin at the shelf edge that will eventually form the base. Again, we want to start with a tight pack for a good crisp bottom edge.

The first pack is complete. Note that we've left an organic edge along the ¾ inch edges of the front panel. You don't want to force a super straight edge that will appear as an artificial seam later.

At this point the concrete is only 1/2" to 3/4" thick at the larger lumps. It needs to cure for an hour or so. This gives you time to measure and cut wire mesh reinforcement for application at the next stage.

A new mix is created. You want a really stiff, peanut butter-like texture. This is then smoothed on, buttering the inside to make it smooth, with a thin application.

This time you should add polypropylene DuraFibre™, one big handful per bag. Fiber is not used on the surface areas of the casting because then you end up with a peach fuzz effect. You can burn the fuzz off with a blowtorch, or sometimes they sand out. But mostly the fibers just get in the way.

A creamy texture is desirable.

A thin layer of the new mix is buttered on to create a smooth surface.

Add reinforcement wire to the smoothed inner surface. Add corner reinforcement first. It's important that there be one piece going through the corners to give them strength.

Wet mix is thrown up on top of the wire, and the 6" putty knife is used to impregnate the wire with the mix, getting all the air out.

This is the perfect time to put an old putty knife with dings and scratches to work.

The top of the bar area is reinforced, here on the bottom of the mold.

The side wall of the center unit is reinforced.

All of the areas have been double reinforced with wire mesh, and impregnated with concrete mix.

Build up the edges on the sides of the front panel to be attached.

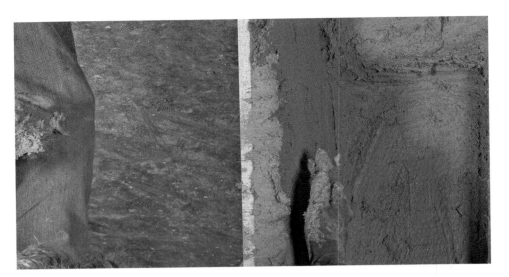

Also, build up the edges on the upright form where they will meet the front panel.

Get a purchase on the base of the front panel using putty knives.

Prop the panel up on a 2' x 4' board and ease to the edge of the upright mold.

TOP LEFT Raise the horizontal panel up to meet the upright mold and align it.

TOP RIGHT Clamp the base and the top of the mold.

LEFT After the clamps have been secured, screw the panel on using the predrilled holes.

More wet mud is used to finish off the bottom to the 2" lip to create a solid base for the unit.

A 1' x 4' board is used to integrate the corners, drawing the mix from the bottom up. The finished casting needs to cure for 12 hours, or overnight.

CASTING THE COUNTERTOP WITH INTEGRAL SINK & DRAINBOARD

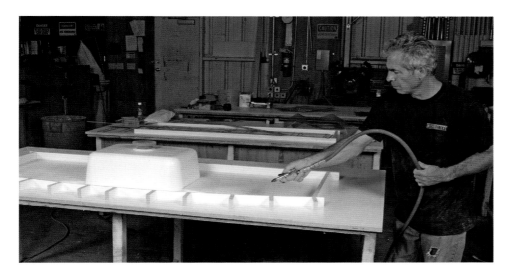

Use a blower to ensure that the mold is dust free.

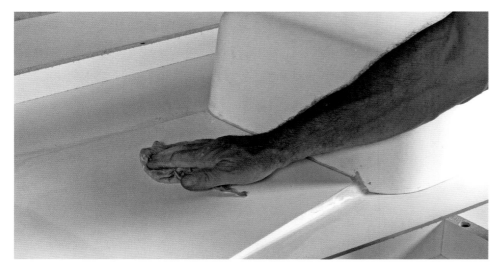

Use a damp rag for a final clean.

Apply a clear paste wax to the sink mold to act as a release.

Mix the first batch of concrete as illustrated in center unit mold pour, substituting Chocolate pigment. Start the application of Chocolate colored mix in the rear sink area, pressing the mix firmly into the corners.

Work evenly over the surface of the mold, and build up the sink from the base.

Take care in working the mix into the angles of the integral drainboard.

Work the mix firmly under the lip of the drain knockout.

Once all the surfaces are covered, follow with a firm, flat-handed smoothing, pressing the surfaces evenly all over.

IMPORTANT: Let sit for about 45 minutes.

This wetter mix is about 4.5 quarts of water per 70-lb. bag. Add a handfull of polypropylene fibers to the mix water before mixing, to give more reinforcement strength.

Use this batch to butter the first layer to accept the wire reinforcement.

Fit the Ladur™ wire reinforcement carefully into the mold, making sure it lies level and doesn't come too close to the edge of the mold.

After all the flat Ladur™ wire is in place, add the first layer of mesh to the sink mold…

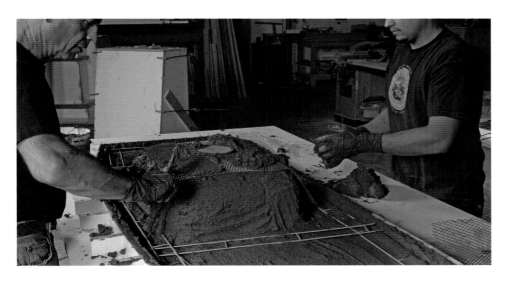

… and butter with wet mix to secure it.

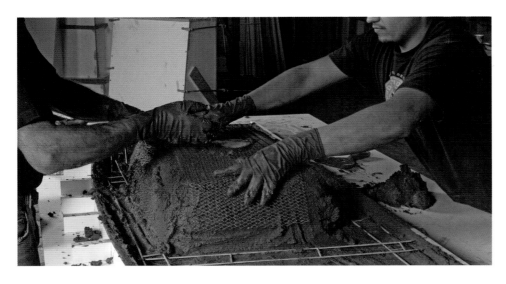

Add the second layer of mesh and secure with mix.

Add Ladur™ wire reinforcement to sink area.

Secure reinforcement with concrete…

… and smooth.

Add final layer of reinforcement

Fill mold and screed flat.

Use a nail to test the depth of the mix on the sides of the sink mold.

More mix should be added to thin areas.

Use a wood float to smooth the sink area.

Use the custom screed to ensure a 1-1/2″ thickness on the backside of the sink.

Make final passes with the wood float to create a pleasing appearance for the underside of the sink, which will show beneath the counter.

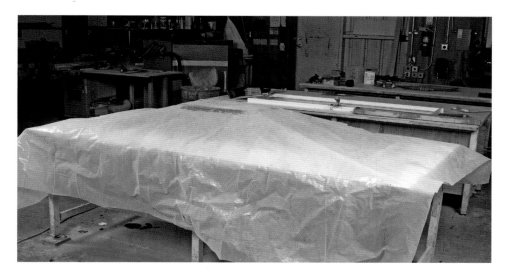

Cover with plastic and allow to cure for 12 hours.

TAKING OFF THE MOLD

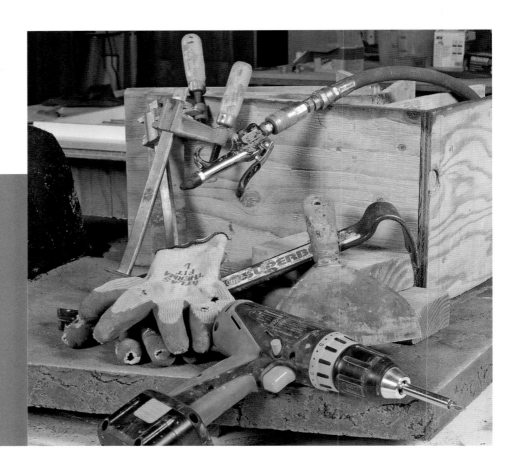

At right are detail shots of the custom supports I make from 1/2" plywood and 2' x 4' boards to hold up the countertop with integral sink.

Remove the screw holding the drain knock-out in place.

Remove the screw holding the faucet knockout in place.

Remove the screws holding the mold walls, and the radius edge block supports.

The mold walls should come away easily. Remove the two screws holding the sink mold to the base.

Raise the sink by first inserting a 6-inch putty knife and …

… applying pressure with a pry bar.

Progressively larger blocks of wood on either side will raise the counter up so you can get a grip on it to turn it. You need to have blocks or angled braces on hand for the next step.

Tipping the counter onto its side is a two-man job. The white paint, visible as a result of our painting the bonds in the mold, will come off easily during polishing.

After the sink is securely supported on blocks, it's time to remove the sink mold. There are three ways to do this: you can roll it over and use an air hose underneath, against the hole created by the drain knock-out; you can fill it with ice to shrink the fiberglass, or you can use clamps.

To use clamps, build a .bridge support over the sink mold and insert clamps into the opening at the top of the sink mold. Apply equal pressure to the clamps to pop it free of the concrete.

Or use the air hose in the hole created by the drain plug.

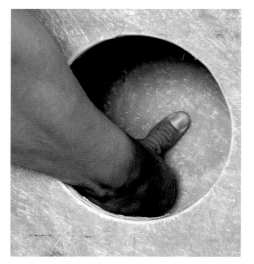

Put a finger on the inside hole, forcing the air between the mold and the sink walls.

The sink is free of the mold.

REMOVING THE MOLD ON VERTICAL CASTING

Remove the clamps.

Remove screws securing the front panel.

The front panel comes away easily.

Remove the top boards creating the 2″ lip for the base.

Use a 6″ putty knife to scrape away excess from base, and use brick stone to grind smooth.

Place a floor pad in front of the casting and carefully lay the piece on its face.

Remove the shelf knockout and the top panel.

The rest of the mold should come away easily.

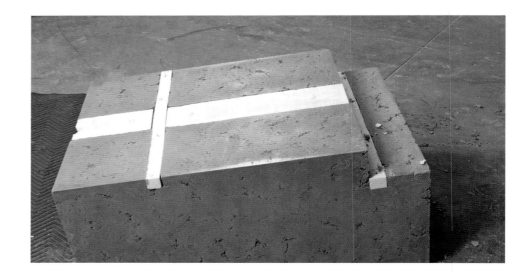

Only the foam knockouts remain.

Use putty knives and pry bars to create room for a grip in order to elevate the casting.

The casting is raised.

The foam knockouts come out easily.

A little prying convinces the more reluctant central gullet knockout.

The joined edges have some overhang; use 6" putty knife to cut away overhanging pieces.

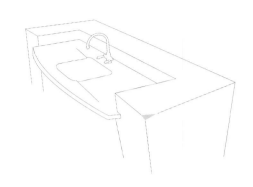

FIXING IMPERFECTIONS

MOLD FAILURE, which can cause cracks and imperfections, is not uncommon. The beauty of this system is that it's so forgiving.

One corner is imperfect because the melamine cracked during the attachment of the front panel. We'll fix that.

The corresponding crack in a corner of the melamine panel.

Use a hammer …

… and chisel to chip away the bad area of the corner.

Use the brick stone to smooth the corners.

An air gun removes the dust.

Spray water on the seam areas to make them more receptive to patching.

Combine Acrylic Additive and water in equal portions.

Add a couple drops of color. The amount will be judged by eyeball, and we don't need very much.

Add a couple handfuls of concrete mix, and mix until combined.

MATCH your patching material color with wet, not dry, concrete.

PROTIP

Add more mix until a doughy consistency is achieved, and color until the mix matches the wet portions of the casting. This can be measured mathematically, using division and scales to reduce the formula from the standard two-bag blend.

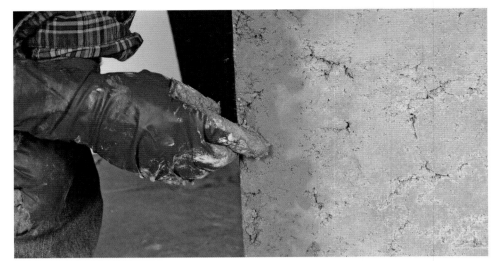

Spray water on areas to be fixed for better bonding. Use the dough-like mix to fill the big voids, but not everything. We want it to retain that rustic look.

Press wet mix into the seam.

Use small, circular sweeps with the straight edge 6" putty knife to knock the wet mix flat. Allow to cure for 12 or more hours.

CURING & FINISHING

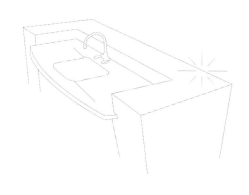

BEFORE WE START our finishing methods, we will dry-fit the pieces of our outdoor kitchen unit. Once we are confident that we have a working combination, we will move forward into pasting, polishing, and sealing.

After the form is removed, the pieces are assembled. Here we see the sink area from the front.

The bottom shelf is dry fit to make sure if fits.

Check to make sure the knock-out slots align. A level assures the fit.

Six bags of mix were used to make the counter with integral sink, and two bags for the bottom shelf. The counter with integral sink weighs about 280 pounds, making moving it a three-man job. Fitting it into the slot takes a little shimmying, but it's a perfect match.

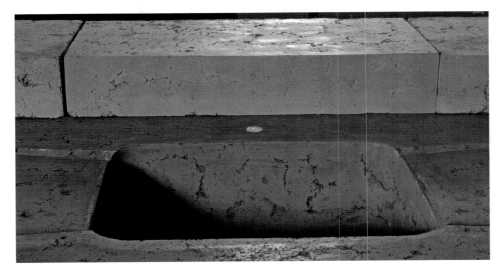

A close-up reveals the unfinished colors and textures of the newly revealed casting before finishing.

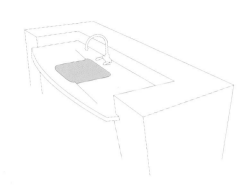

LAMINATING THE SINK

APPLY A LAMINATING POLYESTER RESIN to the underside of the sink. It's assurance that, if the sink cracks, it will prevent leakage. This is just another layer of protection.

Use a brick stone to knock off nubs and large chunks, and grind down the edges on the bottom.

Follow the manufacturer's instructions for measuring and mixing the resin and catalyst, then add a big pinch of polypropylene fibers to strengthen the resin coat.

SAFETYTIP RESIN emits strong fumes. It is important to wear a respirator, and work outside if possible.

Use a disposable brush and apply liberally, spreading the mixture evenly.

The resin will require several hours to cure completely.

FILLING THE VOIDS

I ALWAYS LIKE to leave voids, to provide texture and depth when I can. Of course, we mustn't leave holes in the coutertop and shelf where food will be prepared and displayed. On the unit holding the countertop, however, we're freer. We will fill all the voids in the horizontal slabs and polish them smooth, leaving nooks and crannies in the vertical unit for a rustic look.

Grind the edges with a diamond cup wheel on a 4-1/2" grinder. The purpose is to ease the edges so they aren't so crisp. At present they are vulnerable to chipping.

Make sure to knock all the edges off.

A close-up of the ground edges.

Blow the dust off. This is important for the paste to stick properly.

Mix color paste using acrylic additive. This helps it bond better and set up more quickly. I'll be using three colors on the wheat base. First Sand, then Mushroom and finally Ash, which is darker. Follow the manufacturer's recommendations.

You should end up with a paste of yogurt consistency.

Apply the paste using latex gloves. Don't try to fill in all the holes. As this is the first color, the other colors will provide more coverage.

I am trying for a rustic look in the finished piece, which means that I want some voids in the finish. Be sure to cover the complete surface. If you don't it will become quite apparent at the polishing stage.

When the first color is dry apply the second in the same manner.

Mixing the mushroom tone.

Again, I try not to fill all the voids. This coat should go on more smoothly since the first lubricated the surface.

Wait until the second coat is dry, about ½ hour, and apply the third. I am mixing the ash color. This mixture is slightly more watery.

In applying the final coat, I am skimming it more than the other colors, covering the surface, but not trying to work it into the voids.

When the colors are all applied the pieces should be left to cure for four to 24 hours. Several factors contribute to the length of time, including heat and humidity. If the color sets too long it will be harder to remove in the polishing process, and will allow you to get a smoother finish. We're going for a rustic look, so a long cure isn't necessary.

FILLING VOIDS IN THE COUNTERTOP AND SHELF

Likewise, ease the edges of the countertop, working around the outside …

… and the sink. Don't grind the edge of the drain depression.

Remove the dust.

Mix the first color slip using Chocolate.

Test the color first; match infill paste color to that of the *wet* countertop. This is too light.

PROTIP

YOU CAN TWEAK the powdered paste with a little of the Liquid Color.

If it's too light, add some liquid pigment. Then, transfer the slip to a tray.

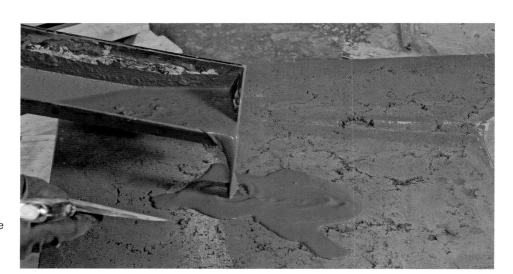

Pour small quantities of the slip onto the countertop surface.

You want a smooth surface, so apply the slip with a 6" putty knife, filling the voids. The edge of the putty knife should be very clean with no nicks or bends. I replace this tool often with new, using the older one for cleaning molds or other less critical tasks.

In the curved sink a potter's "kidney" tool works the slip into the corner.

A hand makes a good tool, as well.

Remove the excess slurry and let it dry.

The second round of color is slightly darker, using a bit more of the liquid pigment. This gives some interest to the final finish.

Apply the second coat with the putty knife and work the excess into the voids. Work to create a smooth finish on the sink work area.

Let it dry. Because we want a smoother finish, this piece should cure longer.

After curing for several hours, repeat the filling process until all the voids are full.

Another coat fills voids created by shrinkage. Some deeper voids take several applications. Polishing will reveal some pinholes, which we will seal and paste again.

POLISHING: GRIND, GRIND, AND GRIND SOME MORE

TO BE REALISTIC, we should include page after page of grinding images, since in the course of making a concrete countertop, grinding is where the bulk of your time will be spent. It's a monotonous task, and we found we could only show so many images without being repetitive. However, as you bend to your task, keep in mind that you'll be grinding for a long time.

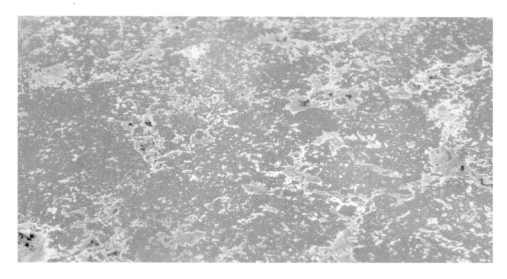

A close-up shows how grinding has removed the grey slip from already smooth surfaces of the cast, while highlighting the voids.

Wet grinding is a messy job. We advise that you wear raingear to protect your clothing. Start off with wet grinding using a 400-grit diamond pad for a light polish on the vertical units.

Turn the pieces on their sides to give your-self horizontal surfaces to work on, which is much easier than sanding a vertical sur-face. Be sure to cushion your piece with a 2' x 4' when you gently lower it, making it that much easier to get under it for moving again later and to prevent chipping from contact with a hard surface.

Grind, grind, grind.

You'll need to hand sand the more difficult and delicate areas of the cast. You could sand the whole thing, but it would take a lot of elbow grease.

A corner unit in the foreground shows the result of an hour's effort at grinding; the other two pieces are waiting in line.

Wet grind the sink countertop.

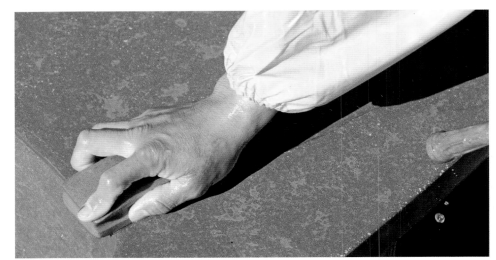

You need to hand sand the rounded and more difficult areas.

The bottom shelf gets the same wet grind treatment, both turned on its side...

... and lying flat.

INITIAL SEALING

Spray on your first application of penetrating sealer. When it's brushed on, the first brush stroke tends to show – who knows why – but it gives a dark brush impression. With the more haphazard spray application we avoid that.

After spraying evenly across the surface and sides, spread the penetrating sealer evenly using a soft cloth. Allow to cure for two to three hours.

The same process is undertaken with the bottom shelf; a soft cloth finishes the process.

After allowing penetrating sealer to cure for a couple of hours, apply satin sealer with a soft cloth. This is a sacrificial sealer, as it will be lost when you refill the small remaining voids and grind again. However, it will help to protect the surface during the process.

Add a final layer of paste to the integral sink and counter, smoothed on with latex gloves. The goal here is to fill the tiny voids and holes and create a smooth work surface for the kitchen area.

Do the same for the bottom shelf.

Grind some more. Use the wet grinder and work the entire surface smooth.

Hand sand the more difficult edges and corners.

FINAL SEALER ON COUNTER
WITH INTEGRAL SINK

BECAUSE WE WANT the countertop and shelf area to have a finely ground sheen, and to keep it through all weather conditions, we chose to finish it with three coats of lacquer. Each was carefully applied, then polished with a Scotchbrite™ pad. Before actually using the countertop, we'll coat it with our food-grade wax.

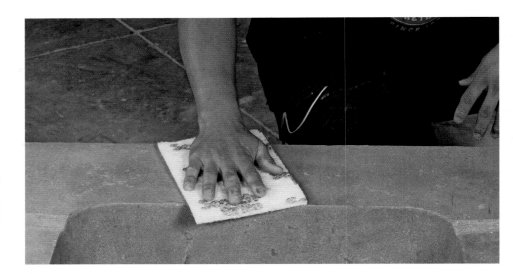

After the final slip application dries there will be some sludgy dust. Go over the surface carefully with a Scotchbrite™ pad to remove all the dust. Use a blower and soft rag for a final once over before sealing.

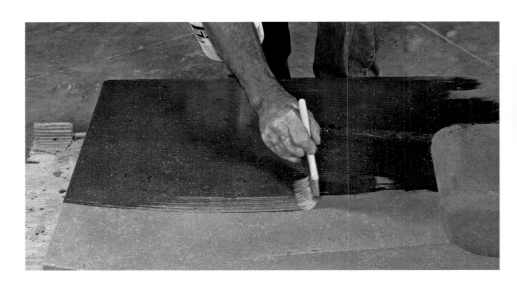

Apply lacquer with a good brush. Take care as the brush strokes will show. It is a good idea to wear that respirator again.

After about half an hour, use a Scotch-brite™ pad to knock it off again. Repeat the same process on the bottom shelf. Apply multiple coats for sturdy outdoor use.

Push out the faucet knockout.

The same process will be repeated on the vertical support units. However, we'll only need one application of lacquer because we want these pieces to be more rustic in contrast to the work surfaces.

ONCE THE SEALANTS DRY, replace the unit's counter and shelf components. The piece is now ready for installation and use.

GALLERY OF
IDEAS

THE IMAGES IN THIS GALLERY are examples of finished concrete projects made with relatively complex three-dimensional molds, radius edges, and vertical concrete application. The Outdoor Kitchen detailed in this book uses similar methods to those necessary for the pictured projects.

Curved tiles attached to fiberglass columns. The Sephora at the Venetian Hotel, Las Vegas. The effect was created with a pressed finish in Bone and Ash colors. Cast with multiple curved edge tile forms on flat tables.

PHOTO BY KEN GUTMAKER

Surround and Hearth were created with a hard-trowelled finish; Universe color Constructed in multiple three dimensional molds built from a design by fireplace designer Jim Scott of Buddy Rhodes Studio.

DAVID DUNCAN LIVINGSTON PHOTOGRAPHY

Curved Surround and Raised Hearth. Hard trowelled finish; moss color. Three-dimensional molds built from drawings by fireplace designer Jim Scott of Buddy Rhodes Studio.

DAVID DUNCAN LIVINGSTON PHOTOGRAPHY

Fireplace surround in pressed finish, moss color. Cast in three-dimensional molds made from design by LouAnn Bauer.

DAVID DUNCAN LIVINGSTON PHOTOGRAPHY

ABOVE Fireplace surround with hard trowelled finish in Universe color. Cast in three-dimensional molds built from design by Jim Scott of Buddy Rhodes Studio.

DAVID DUNCAN LIVINGSTON PHOTOGRAPHY

TOP RIGHT Fireplace surround; hard trowelled finish; custom color. Cast in three-dimensional molds from design by Jim Scott of Buddy Rhodes Studio for designer Jay Jeffers.

DAVID DUNCAN LIVINGSTON PHOTOGRAPHY

LEFT Three dimensional angled fireplace surround; universe color, hard trowelled finish. Design by Larry Strunk.

DAVID DUNCAN LIVINGSTON PHOTOGRAPHY

Drum Stool; Buddy Rhodes Design. Cast in three dimensional fiberglass mold; Pressed finish. Earth color with Ash and Smoke infill.

DAVID DUNCAN LIVINGSTON PHOTOGRAPHY

Cone table designed by Buddy Rhodes. Kitchen design by Glenn Dugas. Cone pressed in three-dimensional mold; ash with coal infill colors.

PHOTO BY KEN GUTMAKER

RIGHT Kidney Shaped Island with Counters. Hard trowelled; Universe color; Curved edge mold.

Outdoor Kitchen countertops in the Wine Country with integral sink. Pizza oven roof cast in three-dimensional mold. Paul Bertolli design.

42 Degree Café Patio Table, Benches, Sphere Planters

ABOVE Buddy Rhodes Lounge Chair, McEnvoy Ranch

PHOTO BY KAREN THOMPSON

RIGHT Sphere planter in Washington State.

PHOTO BY MICHAEL JENSEN

ABOVE The Buddy Rhodes Concrete Product line, with which these projects and many others have been crafted.

For a listing of national distributors, visit **www.buddyrhodes.com**.

RIGHT A glossary of terms, more detailed instruction, and descriptions for the complete Buddy Rhodes product line can be found in his companion book, *Making Concrete Countertops with Buddy Rhodes*. This volume offers photographs and explanations of beginner- to intermediate-level techniques, and professional tips from Buddy Rhodes' many years of experience. Like this advanced-level book, it includes a gallery of ideas to encourage those new to concrete to think bravely and artistically about the material's potential. Pick it up at your local bookstore, or see page 2 of this volume for ordering information.

LOOKING FOR TOOLS?
Visit **www.concretecountertopspecialties.com** to find specialty tools, such as a water-fed hand-polisher like the one used on pages 125-131.

Making Concrete Countertops with Buddy Rhodes (2007)
ISBN No. 978-0-7643-2477-2

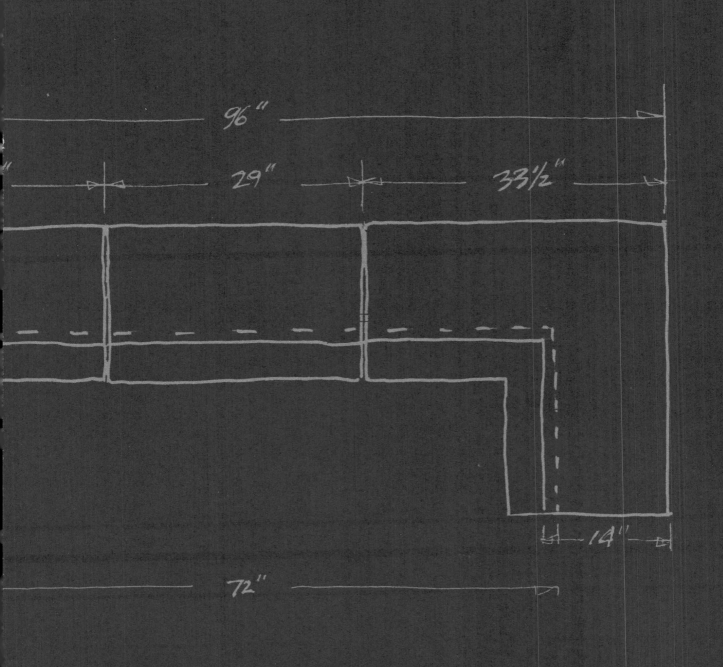